Humanity's Ledger

Humanity's Ledger

The Trust Protocol

Aaron Vick

AaronVick

CONTENTS

Preface 4

Chapter 1: Introduction to
Blockchain and Human Trust 6

Chapter 2: The Evolution of Trust
in the Digital Age 17

Chapter 3: Blockchain: Beyond
Transactions 27

Chapter 4: Humanizing
Technology: The Blockchain Paradox 35

Chapter 5: Work and Purpose in
the Age of Automation and
Blockchain 50

Chapter 6: Redefining Societal
Contracts in the Blockchain World 66

CONTENTS

| Chapter 7: Trust in a Decentralized
World: Opportunities and Challenges   75

| Chapter 8: Future Visions:
Humanity and Technology in
Harmony   88

| Chapter 9: The Path Forward   94

References   102

Additional Reading   107

Preface

In a world rapidly reshaped by digital innovation, this narrative traverses the intricate relationship between blockchain technology and the multifaceted concept of human trust. This exploration ventures beyond the conventional discourse on cryptocurrencies and NFTs, delving into the philosophical and sociological implications of blockchain in our social fabric.

Will we allow technologies to constrain the richness of humanity, or can we direct them to liberate our creative potential? Can math-driven systems strengthen, not sever, the bonds between us? This narrative explores a future where digital platforms amplify, rather than diminish, the four key attributes that make us uniquely human - our capability for empathy, creativity, humility and trust.

Today, much discourse focuses on blockchain as merely a better database - an efficient way to track assets and transactions. But at its core, blockchain has philosophical implications. Beyond enhancing ledgers, it can enhance human character and community. Beyond tracking packages, it can foster understanding. Ledger is ledger, but life is life.

Technology is never destiny - it emerges from human choices, not abstract forces. We must guide blockchain's evolution with conscious intention, ensuring its arcs bend towards justice. Applications must uplift human dignity, unlock expression and enlarge circles of compassion. Systems claiming immutability must make room for people to mature.

Yes, blockchain enables certainty - but humans often thrive in uncertainty's fertile soil. Structure enables security, but communities flower through forgiveness. Perfection is an illusion, but human potential is boundless. With care, blockchain's resolute trust can resonate with relationships' ever-changing rhythms, harmonizing our headlong rush into the digital future with timeless wisdom.

Progress takes patience; revolutions arise through millions of tiny acts of love. A better world is created brick by blockchain brick. The technology is ready - but are we ready to guide it humanely? The plot has not yet been written. We grip the pen; let us co-author this story together.

Chapter 1: Introduction to Blockchain and Human Trust

"Trust is the glue of life. It's the most essential ingredient in effective communication. It's the foundational principle that holds all relationships." — Stephen R. Covey

---

In the midst of our rapidly changing digital world, a perplexing contradiction brews – the pursuit of trust, an essential ingredient of human relationships, is now intertwining with the groundbreaking technology of blockchain. This chapter embarks on an enlightening journey, unraveling the intricate connection between blockchain technology and the timeless human concept of trust.

The Dawn of a New Era in Digital Trust

As we advance into uncharted territories of the digital age, blockchain technology emerges as a beacon, heralding a transformative era that redefines the essence of trust in our digital interactions. Initially the backbone of cryptocurrencies like Bitcoin, blockchain has burgeoned into a multifaceted force, with implications extending far beyond its financial origins. It challenges our traditional understanding of trust, security, and transparency, presenting an alternative to the centralized systems that have long underpinned our societies.

This technology, at its inception, was more than a mere advancement; it was a philosophical and sociological paradigm shift. It invites us to ponder: How does this technology, grounded in mathematics and cryptography, align with the dynamic and often elusive nature of human trust? Can the digital certainty of blockchain harmonize with the intricate, emotional fabric of trust that connects us as humans?

Delving into Blockchain's Core Principles

In this section, we explore the foundational pillars of blockchain technology – decentralization, immutability, and transparency – and their groundbreaking role in establishing a new paradigm of trust in the digital world. We examine how blockchain's decentralized nature eliminates the need for central authorities, how its immutable ledger offers an unalterable record of truth, and how its transparent nature fosters unprecedented levels of confidence and reliability in our digital interactions.

The Human Dimension of Trust in a Digital Context

Shifting focus to the realm of human societies, we delve into the bedrock of relationships and communities - trust. This section navigates the multifaceted landscape of human trust, an entity that is dynamic, emotional, and deeply embedded in personal interactions and societal norms. We juxtapose the binary certainty of blockchain against the nuanced, complex shades of human trust, exploring how historical reputation and societal contracts shape our perception and experience of trust.

Bridging Two Worlds: Beyond Mere Transactions

As we venture further, we confront the intricate task of blending blockchain's digital trust with the multifaceted nature of human trust. This section contemplates blockchain's formidable capability in securing transactions and records while acknowledging its limitations in encapsulating the qualitative aspects of human relationships. It calls for an innovative integration of blockchain technology with the social and emotional dimensions of human existence, envisioning applications that transcend mere transactions to foster community building, social engagement, and empathetic understanding.

Charting a Path Towards a Synergistic Future

Navigating this complex interplay, the chapter concludes with a vision of a future where blockchain and human trust coexist in a harmonious synergy. This journey challenges us to innovate in ways that uphold our humanity while leveraging the transformative power of blockchain. By embracing this challenge, we envisage a digital world that is secure, efficient, and imbued with compassion, adaptability, and a deep sense of human connection.

Overview of Blockchain Technology and Its Fundamental Principles

Blockchain technology, since its inception, has revolutionized the way we perceive data security and trust in the digital realm. Originally conceived as the underlying structure for cryptocurrencies like Bitcoin, blockchain has evolved to have far-reaching implications across various sectors.

Key Characteristics of Blockchain:

- **Decentralization**: Unlike traditional centralized systems where a single entity has control, blockchain operates on a decentralized network. This means that no single individual or organization holds the authority over the entire network, making it inherently resistant to centralized points of failure and control.
- **Immutability**: Once data is recorded on a blockchain, it becomes nearly impossible to alter. This immutability is ensured through cryptographic hashes, a unique string of numbers and letters representing the data. Each block contains its own hash and the hash of the previous block, creating a linked chain that secures the entire history of transactions.

- **Transparency**: Blockchain's ledger is open for anyone to verify and audit, fostering a level of transparency uncommon in many traditional systems. This feature is pivotal in building trust among users who can independently verify the integrity of the data stored.
- **Consensus Mechanisms**: Blockchain employs various consensus mechanisms like Proof of Work (PoW) and Proof of Stake (PoS) to validate new transactions. These mechanisms ensure that all participants in the network agree on the validity of transactions, preventing fraud and duplicity.

Blockchain's Impact on Trust:

The principles of blockchain introduce a new paradigm of trust. In traditional systems, trust is often established through intermediaries like banks or governing bodies. Blockchain, however, enables trust to be established directly between individuals, thanks to its decentralized, transparent, and immutable nature. This direct trust negates the need for intermediaries, reducing costs and increasing efficiency.

In a blockchain network, trust is not placed in a single entity but distributed across a network of nodes, each verifying the transactions independently. This distributed trust model is a significant shift from the centralized trust models that have dominated societies for centuries.

The integration of blockchain into various industries shows its potential to enhance transparency, security, and efficiency. From supply chain management to voting systems, blockchain's ability to ensure the integrity and traceability of data is transforming traditional operations.

Exploration of the Concept of Trust in Human Societies

In human societies, trust serves as the bedrock upon which relationships and communities are built. It's a nuanced, dynamic concept, essential to the functioning of social structures, economies, and interpersonal connections. This section delves into the complexities of human trust, setting the stage for understanding how blockchain technology intersects and contrasts with these traditional trust paradigms.

The Multifaceted Nature of Human Trust:

Trust among humans is deeply emotional and relational. It evolves through personal interactions, shared experiences, and mutual understanding. This emotional dimension of trust, integral to human relationships, stands in stark contrast to the logical, binary nature of blockchain systems. Human trust is not static; it's a dynamic entity, shaped and influenced by a myriad of factors including cultural norms, past experiences, and personal values. In human societies, trust often resides in shades of gray, allowing for nuances and changes over time, unlike the binary trust of blockchain.

Historical reputation plays a significant role in human trust. People and institutions build trust over time through consistent and reliable behavior. This aspect of trust, based on history and reputation, finds a parallel in blockchain technology with its transparent, immutable record of transactions.

Trust in Societal Structures:

Traditional societal trust is often vested in established institutions like governments, banks, and legal systems. These entities act as intermediaries in societal interactions, establishing trust based on rules and regulations. The concept of social contracts is pivotal in societal trust. These are the tacit agreements that members of society adhere to, ensuring cooperation for the common good. When these social contracts are violated, it can lead to a breakdown in trust.

Blockchain's Challenge to Traditional Trust Models:

Blockchain technology introduces a new way of establishing and maintaining trust. By eliminating the need for central intermediaries and offering a transparent, immutable ledger, blockchain presents a significant challenge to traditional models of trust:

Blockchain's decentralized nature shifts the paradigm of trust from centralized authorities to a distributed, consensus-based system. This decentralization is a radical departure from the trust placed in single, often centralized, authoritative entities in traditional systems.

The immutability and transparency of blockchain provide a form of trust that is objective and absolute. Once data is entered into the blockchain, it cannot be altered, creating a permanent and reliable record. This feature of blockchain stands in contrast to the flexible and often subjective nature of human trust.

The reconciliation of blockchain's technological trust with the multifaceted nature of human trust is a complex endeavor. It involves understanding how to integrate the certainty of blockchain with the fluidity of human interactions. The challenge is to design blockchain systems that not only execute agreements but also respect and understand the complexities of human relationships.

Bridging Blockchain and Human Trust: Beyond Transactions

In the realm where blockchain technology and human trust converge, we encounter a landscape rich in potential yet fraught with challenges. This section explores the deeper implications of blending blockchain's digital trust with the multifaceted nature of human trust, focusing on areas beyond mere transactions and emphasizing the nuances of human interaction and connection in a blockchain-enabled world.

The Limitations of Digital Trust in Human Contexts

Blockchain technology, celebrated for its ability to create secure, transparent records, inherently lacks the capacity to capture the qualitative aspects of human relationships. Its strength lies in providing a robust framework for transactional integrity, yet it falls short in addressing the complexities of trust that are rooted in shared human experiences, emotions, and social dynamics. The nuances of empathy, understanding, and emotional intelligence, which form the bedrock of human trust, remain beyond the reach of any digital ledger.

Humanizing Blockchain: Incorporating Social and Emotional Dimensions

To bridge this gap, it becomes essential to explore ways to humanize blockchain technology, integrating it with the social and emotional dimensions that define human interactions. This involves not just leveraging blockchain for its technical prowess but also understanding how it interacts with, influences, and is influenced by human behavior and societal norms. The challenge is to design blockchain systems and applications that recognize and respond to the human aspects of trust, perhaps by integrating user-centric design principles that prioritize empathy and user experience.

Potential Avenues for Human-Centric Blockchain Applications

One potential avenue is the development of blockchain applications that facilitate community building and social engagement. These applications could go beyond mere transactions, fostering environments where trust is built on shared values, collaboration, and mutual support. For example, blockchain could underpin platforms for decentralized social networks, community-driven initiatives, or collaborative projects, where the focus is on building relationships and trust among participants.

Navigating the Paradox: Respecting Human Trust in a Blockchain World

As we navigate this paradoxical landscape, it becomes clear that the future of blockchain in society hinges on our ability to respect and incorporate human trust elements. This requires a holistic approach where technological innovation is not seen in isolation but as part of a larger ecosystem that includes human values, emotions, and social structures. The aim is not to supplant human trust with digital trust but to create a synergistic relationship where each enhances the other.

Toward a Synergistic Future

As we navigate the intricate pathways towards integrating blockchain technology with the nuanced fabric of human trust, we find ourselves on a journey marked by complexity and continuous evolution. This journey compels us to delve deeper into the essence of trust within our increasingly interconnected digital landscape, urging us to forge innovations that not only harness technological prowess but also resonate with the core of our human experience.

In embracing the multifaceted challenges and opportunities presented by this integration, we stand on the cusp of shaping a future where the role of blockchain transcends the mere securing of transactions. We envision a world where blockchain acts as a catalyst, not just for technological advancement but for enriching the very nature of human connections. This future envisages a digital realm infused with compassion and adaptability, mirroring the depth and complexity of human relationships.

In this envisioned world, blockchain becomes more than a tool; it emerges as a bridge connecting the certainty of digital trust with the fluidity of human interactions. It fosters a digital ecosystem where technology and humanity coalesce, creating a space that values security and efficiency while cherishing empathy and understanding.

We are in an era that beckons us to not only contemplate the potential of blockchain but to actively participate in shaping a future where technology enhances, rather than diminishes, the richness of human relationships. In doing so, we pave the way for a digital world that truly reflects the best of both our technological capabilities and our human values.

Chapter 2: The Evolution of Trust in the Digital Age

"The first step in the evolution of ethics is a sense of solidarity with other human beings." — Albert Schweitzer

---

As the digital era unfolds, it brings with it a transformative wave that redefines the ancient concept of trust, a bedrock of human connection and societal cohesion. This chapter delves into the intricate evolution of trust, tracing its journey from the tangible bonds of ancient societies to the complex digital networks of today's world.

Gone are the days when trust was solely the product of face-to-face interactions and shared community experiences. In our current era, trust extends beyond the physical realm, weaving into the fabric of digital interactions. This shift presents a compelling narrative: how does trust, a concept deeply rooted in human emotion and societal norms, adapt and thrive in the digital landscape shaped by technologies like blockchain?

We explore the cultural and societal underpinnings of trust, unraveling how it varies across civilizations and evolves over time. From the communal fires of ancient tribes to the interconnected networks of the modern digital society, trust remains a pivotal force, yet its expression and mechanisms of establishment undergo profound changes.

As we venture through this chapter, we investigate the intersection of trust with digital technologies. This exploration is not just about understanding the impact of these technologies on traditional trust paradigms, but also about envisioning how trust can be reimagined and strengthened in a world increasingly dominated by digital interactions.

Historical Perspective on Trust in Human Relationships

Trust has long been the cornerstone of human relationships, a fundamental element woven into the fabric of societies and cultures worldwide. Historically, trust has been built through face-to-face interactions, personal experiences, and a shared understanding of societal norms and values. These interactions, laden with emotional intelligence and empathy, have defined how communities are formed, how they operate, and how they evolve over time.

The Cultural and Societal Foundations of Trust

The concept of trust varies significantly across different cultures and societies. In some cultures, trust is built over long periods, deeply rooted in shared history and experiences. In others, trust can be established more rapidly, often influenced by societal structures and communal values. This variability reflects the adaptability of trust as a social construct, one that is shaped by the complex interplay of individual relationships, community bonds, and societal norms.

Trust and Social Cohesion: From Ancient Times to Modern Societies

From the times of ancient civilizations to modern societies, trust has played a crucial role in the formation of social cohesion and the functioning of communities. Trust has been essential for cooperation, trade, and the establishment of social order. Philosophers like Aristotle and Confucius spoke of trust as a virtue essential for societal harmony and individual well-being. In more recent times, sociologists like Niklas Luhmann have emphasized trust as a mechanism for reducing social complexity, allowing individuals to navigate an increasingly interconnected world.

The Role of Personal Experience and Emotional Bonds in Trust

Personal experiences and emotional bonds form the bedrock of trust in human relationships. Trust evolves through interactions, shared experiences, and the emotional connections that bind people together. This type of trust is fluid, often changing over time as relationships grow, and is deeply influenced by the capacity for empathy, understanding, and forgiveness. These human-centric aspects of trust are vital for the development of strong, resilient communities.

Exploring the Intersection with Digital Technologies

As we transition into the digital age, the traditional dynamics of trust are being reshaped by the advent of new technologies. This part of the chapter sets the stage for exploring how digital technologies, especially blockchain, are transforming the concept of trust and its role in digital interactions. The focus will be on understanding how the timeless human quest for connection and trust is evolving in the face of technological advancements, leading to new forms of social interaction and community building in the digital realm.

In the next sections, we will delve deeper into the impact of digital technologies on trust and explore blockchain's transformative role in redefining trust in digital interactions.

The Impact of Digital Technologies on the Concept of Trust

The infusion of digital technologies into the fabric of modern life has dramatically reshaped the concept of trust. This transformation has been rapid and far-reaching, affecting not only how trust is established and maintained but also challenging traditional notions of interpersonal and institutional trust.

Digitization and the Disruption of Traditional Trust Mechanisms

The digitization of communication and transactions has introduced new dynamics in the establishment of trust. In the pre-digital era, trust was predominantly built through direct interactions and long-standing relationships. Digital platforms, however, have enabled interactions among strangers, necessitating the development of new mechanisms to establish and verify trust. The rise of e-commerce, social media, and online communities has led to the emergence of reputation systems, peer reviews, and digital endorsements as new proxies for trust.

Technology as a Double-Edged Sword in Trust Formation

Digital technologies offer unparalleled convenience and connectivity, but they also bring challenges to the traditional understanding of trust. The anonymity and distance provided by digital interactions can lead to a decrease in accountability and an increase in fraudulent activities. This erosion of trust is further exacerbated by the prevalence of misinformation, identity theft, and cyber threats in the digital realm. The challenge lies in leveraging technology to enhance trust without compromising the authenticity and integrity of human relationships.

Blockchain as a Game Changer in Digital Trust

Blockchain technology stands out as a revolutionary force

in the realm of digital trust. Its key features – decentralization, transparency, immutability, and cryptographic security – provide a new foundation for establishing trust in digital interactions. Blockchain creates a verifiable and permanent record of transactions, reducing the need for traditional intermediaries and fostering a trust protocol that is both secure and transparent. This technology opens new avenues for creating and maintaining trust in digital societies, from financial transactions to identity verification and beyond.

The Societal Implications of Trust in the Age of Information

As we navigate the digital age, the role of information in trust formation has become increasingly significant. The ability to access, share, and verify information is central to the development of trust in digital platforms. However, the deluge of information and the complexity of discerning its authenticity pose new challenges to trust in the digital landscape. The democratization of information, while empowering, also necessitates a critical examination of its impact on trust, particularly in the context of media, governance, and public discourse.

In this landscape, the role of blockchain emerges as not only a technological solution but also a catalyst for a deeper discussion on the evolution of trust in the digital age. The next part of this chapter will delve into blockchain's transformative role in redefining trust in digital interactions, exploring its potential and limitations in shaping the future of trust in our interconnected world.

Blockchain's Transformative Role in Digital Trust

The advent of blockchain technology has introduced a pivotal shift in the landscape of digital trust. This transformation is not just technological but also deeply cultural, as it redefines the mechanisms and essence of trust in an increasingly digital society.

Blockchain as a Trust Architecture in the Digital Realm

Blockchain technology offers a novel architecture for trust in digital interactions. Its decentralized nature eliminates the need for central authorities or intermediaries, thereby reducing potential points of failure and corruption. This shift from centralized to decentralized trust systems represents a fundamental change in the power dynamics of trust, granting more control and transparency to the users.

Immutability and Transparency: Core Pillars of Digital Trust

The immutability and transparency of blockchain are its most distinguishing features. Once data is recorded on a blockchain, it cannot be altered retroactively without the consensus of the network, providing a tamper-proof record. This feature builds a strong foundation of trust, as every transaction is verifiable and permanent. The transparency of blockchain further enhances trust, as all participants in the network have access to the same information, ensuring a level of openness that was previously difficult to achieve in digital systems.

The Limitations and Challenges of Blockchain in Establishing Trust

While blockchain provides a robust framework for trust, it is not without its limitations. The technology, in its current form, primarily addresses the trust in the integrity of data and transactions. However, it does not automatically solve the challenges of establishing trust in identities or the quality of goods and services transacted over the network. Additionally, the technology's complexity and the need for widespread understanding and adoption pose significant challenges in fully realizing its potential to transform trust.

Integrating Human Elements into Blockchain Systems

A crucial aspect of advancing blockchain's role in trust is the integration of human elements into these systems. Trust is not purely a technical matter; it is deeply rooted in human psychology and social dynamics. As such, blockchain systems must be designed with an understanding of human behavior, encompassing aspects like reputation, ethical considerations, and relational dynamics. This integration is essential to ensure that blockchain systems are not just technically sound but also resonate with the complexities of human trust.

Looking Ahead: Blockchain as a Stepping Stone to Future Trust Paradigms

The journey of blockchain in the realm of trust is just beginning, a mere glimpse of its potential in the broader landscape of digital integration. As we stand at this juncture, blockchain emerges not just as a technological breakthrough but as a precursor to more profound concepts like digital consciousness. This evolving narrative places blockchain at the forefront of a journey towards a deeply interconnected digital existence, urging us to stretch our imaginations beyond its current capabilities.

Envisioning a future where blockchain is integral to digital-human convergence, we see a world where the boundaries between technology and human interaction blur. In this world, blockchain does more than just secure transactions; it becomes a cornerstone in the building of a society where digital and human elements interlace seamlessly. The potential for blockchain to enhance, rather than supplant, human trust in this digital era is not only promising but also essential.

As we navigate this evolving digital landscape, the impact of blockchain on trust is both profound and far-reaching. With its ability to offer a secure, transparent, and decentralized framework, blockchain stands poised to redefine how we understand and practice trust in our digital society. This exploration is not merely about technological advancement; it's a deeper quest to harmonize the digital with the human spirit.

The path ahead requires us to balance the technical prowess

of blockchain with the nuances of human trust. It calls for a thoughtful approach that considers the societal implications and ethical dimensions of this technology. As we continue to develop and refine blockchain, it becomes increasingly important to keep sight of its role in shaping not just a more secure and efficient digital world but also a more connected and human-centric one.

Thus, as we embrace the possibilities that blockchain technology brings, we also accept the responsibility to guide its evolution thoughtfully and ethically. In doing so, we lay the groundwork for a future where trust, in all its complexity, finds a new expression in the digital age, enriching the tapestry of human relationships and society at large.

Chapter 3: Blockchain: Beyond Transactions

"For the first time, we have a technology where the truth can be programmatically established."
— Andreas M. Antonopoulos

As the digital landscape continues to evolve at a break-neck pace, blockchain technology emerges as a multifaceted powerhouse, stretching its influence far beyond the confines of financial transactions. In this chapter, we delve into the expansive and diverse applications of blockchain, revealing its profound impact on social, political, and ethical realms.

Blockchain, often synonymous with cryptocurrencies like Bitcoin, reveals its deeper layers of potential and versatility. This technology, with its decentralized and transparent nature, serves as an ideal foundation for myriad applications requiring reliability, security, and immutability. Here, we uncover the myriad possibilities blockchain offers, challenging the conventional perception of it as solely a financial tool.

One of the most transformative applications of blockchain lies in its ability to enhance trust and transparency in social systems. For instance, blockchain's role in managing and verifying personal identities, ownership records, and even the intricacies of voting systems marks a significant departure from traditional methods. By providing a tamper-proof, transparent record-keeping system, blockchain stands as a beacon against fraud and corruption, fostering trust in both public and private institutions.

The ripple effect of blockchain extends into the political arena, presenting innovative solutions for governance and decision-making. Its application in voting systems, for instance, promises more secure and transparent elections, potentially reducing manipulation and bolstering public trust in democratic processes. Moreover, blockchain's ability to manage public resources efficiently and transparently could transform governmental operations and decision-making, ushering in a new era of governance.

Real-world case studies in sectors like supply chain management and healthcare illuminate blockchain's role in building trust beyond mere transactions. Whether it's tracking the journey of coffee beans or securely managing patient data, blockchain demonstrates its capacity to not only ensure transactional integrity but also to uphold the integrity and transparency of entire systems and processes.

However, the journey of blockchain is not without its challenges and limitations. Issues such as scalability, energy consumption, and the complexity of its implementation invite scrutiny and demand innovative solutions. Additionally, the ethical implications surrounding privacy and potential misuse are critical considerations in the responsible development of blockchain technology.

As we explore the depth and breadth of blockchain's capabilities beyond financial transactions, we stand on the brink of a major shift in how we interact, govern, and make decisions in our increasingly digital society. This chapter invites us to rethink the potential of blockchain to foster trust, transparency, and efficiency across various domains, presenting an opportunity to reimagine our social and political fabric for the digital age.

Exploring Blockchain's Versatile Applications

Blockchain technology, often associated primarily with financial transactions, possesses a vast array of capabilities that extend far beyond the realm of economics. This section delves into the multifaceted nature of blockchain and its profound impact on various social, political, and ethical contexts.

Blockchain's Multidimensionality: More Than Just Financial Transactions

Initially conceived as the underlying technology for cryptocurrencies like Bitcoin, blockchain's potential applications are diverse and far-reaching. Its decentralized and transparent nature makes it an ideal foundation for applications that require reliability, security, and immutability. These characteristics open up a plethora of possibilities across different sectors, challenging the conventional understanding of blockchain as merely a financial tool.

Social Implications: Building Trust and Transparency

One of the most significant applications of blockchain outside finance is in enhancing trust and transparency in social systems. For instance, blockchain can revolutionize how we manage and verify personal identities, ownership records, and even voting systems. By providing a tamper-proof and transparent record-keeping system, blockchain has the potential to significantly reduce fraud and corruption, thereby fostering trust in public and private institutions.

Political and Ethical Contexts: Shaping Governance and Decision-Making

Blockchain's impact extends into the political sphere, offering innovative solutions for governance and decision-making processes. Its application in voting systems, for instance, could ensure more secure and transparent elections, reducing the risk of manipulation and increasing public trust in electoral outcomes. Furthermore, blockchain can be instrumental in managing public resources more efficiently and transparently, potentially transforming how governments operate and make decisions.

Exploring Blockchain's Limitations and Challenges

While blockchain offers immense possibilities, it is not a panacea for all societal challenges. The technology's limitations, such as scalability issues, energy consumption, and the complexity of implementation, must be acknowledged and addressed. Moreover, the ethical considerations surrounding blockchain, such as privacy concerns and the potential for misuse, require careful consideration and responsible development.

As we delve deeper into blockchain's capabilities beyond financial transactions, it becomes evident that this technology holds the key to reshaping various aspects of our social and political fabric. Its potential to foster trust, transparency, and efficiency across different domains presents an opportunity to reimagine how we interact, govern, and make decisions in a digital society. However, realizing this potential fully requires a nuanced understanding of both the technology's strengths and its limitations.

Blockchain technology, often associated with cryptocurrencies and finance, has found a novel and impactful application in the agricultural sector. This case study delves into its role in transforming supply chain transparency, particularly in the journey of coffee beans from farm to cup. It highlights the practical use of blockchain in enhancing ethical practices and trust in agriculture, providing a tangible example of how this technology can extend beyond its traditional boundaries to benefit other crucial aspects of our society.

Case Study: Revolutionizing Coffee Bean Supply Chain Transparency with Blockchain

In the world of coffee production, blockchain technology has emerged as a game-changer. A pioneering initiative by companies like Bext360 and IBM showcases this transformation. They have implemented blockchain systems to trace coffee beans from their source to the consumer's cup. This blockchain-enabled traceability creates an immutable and transparent ledger of the bean's journey, providing an unprecedented level of insight into each step of the supply chain.

From Farm to Cup: A Journey of Ethical Assurance

In the realm of coffee production, blockchain technology has introduced a transformative approach to tracing the journey of coffee beans, ensuring ethical and sustainable practices from farm to cup. This journey begins at the farm, where each batch of coffee beans is meticulously tagged and its details securely recorded on the blockchain. As the beans make their way through various stages – from farmers to processors, distributors, and finally to retailers – each step in their journey is logged on this immutable digital ledger.

This blockchain-enabled traceability offers a comprehensive, real-time view of the coffee beans' path. It serves as a guarantee of authenticity and quality, ensuring that every hand involved in the process, particularly the farmers, receives fair compensation. By making the entire supply chain transparent, this system not only holds all parties accountable but also assures consumers of the ethical sourcing of their coffee.

Furthermore, this method verifies that the beans are cultivated using sustainable farming practices. Such visibility into the supply chain cultivates a culture of responsibility and trust, bridging the gap between producers and consumers. It reinforces consumer confidence in the product, knowing that their purchase supports both ethical farming and fair trade practices.

This case study exemplifies how blockchain can be more than just a technological tool; it can act as a catalyst for positive change, fostering transparency, equity, and sustainability in global agricultural practices. As we explore the multifaceted impacts of blockchain, this application in the coffee industry stands out as a testament to the technology's potential in enhancing not just economic transactions, but also ethical and environmental standards across industries.

Empowering Consumers and Influencing Markets

The impact of this blockchain application extends beyond the logistical aspects of the supply chain. As consumers become increasingly conscientious about the origins and ethical standards of their purchases, this transparent journey of coffee beans empowers them to make informed decisions. They can verify the ethical credentials of their coffee, from fair labor practices to environmental sustainability. This empowerment influences market trends, as consumers increasingly favor products that align with their values, driving a shift towards more ethical and sustainable practices in the industry.

Wider Societal and Ethical Implications

The application of blockchain in the coffee supply chain is a microcosm of its potential societal impact. It demonstrates how technology can be leveraged not just for efficiency and transparency but as a tool for ethical assurance and trust-building in consumer markets. This model, highlighting the interconnectedness of technology, ethics, and consumer behavior, provides valuable insights into how blockchain can be harnessed to foster a more ethical and accountable global marketplace.

In conclusion, the case of blockchain in coffee bean supply chains exemplifies the transformative potential of technology in redefining trust and promoting ethical practices in the agricultural sector. It stands as a testament to the power of blockchain in bridging the gap between digital innovation and societal values, offering a template for how technology can be harnessed to enhance not just efficiency but also the ethical dimensions of our interconnected world.

Chapter 4: Humanizing Technology: The Blockchain Paradox

"Technology is best when it brings people together."
— Matt Mullenweg

Exploring the blockchain paradox requires a deep dive into the juxtaposition of this steadfast technology with the mutable nature of human society. This section unravels how the immutable and precise characteristics of blockchain contrast with the fluid dynamics of human relationships and societal trust.

The narrative probes the challenge of integrating a technology known for its rigid certainty into the intricate tapestry of human interactions, where flexibility, empathy, and understanding are essential. It delves into the complexities and potential pitfalls of applying blockchain's definitive structures to social contexts that thrive on adaptability and change.

Central to this exploration are scenarios where the unyielding nature of blockchain encounters various aspects of human society, including governance, ethical dilemmas, and personal relationships. The discussion centers on how blockchain, while offering security and transparency, must adapt to the evolving demands of human interaction without losing its core attributes.

The aim is to envision a future where blockchain technology enriches human experiences, transcending its role as a mere efficiency tool to become a facilitator of deeper human connections. The focus is on conceptualizing blockchain as a complementary force that aligns with human values, balancing the benefits of technological innovation with the subtleties of human trust and social norms.

Bridging Blockchain with Human Fallibility

In exploring the paradox of blockchain technology within the context of human society, we encounter a fundamental dissonance: the infallible, immutable nature of blockchain stands in stark contrast to the inherently fallible, ever-evolving nature of human interaction and trust. This section delves into the complexities of reconciling these two seemingly disparate realms, highlighting the need for a nuanced approach that recognizes the limitations and potential of both.

Blockchain, at its core, is a technology of absolutes. Its cryptographic foundation ensures a level of security and trustworthiness that is unprecedented in digital transactions. Transactions, once recorded on the blockchain, become an unalterable part of a permanent ledger, offering a level of certainty and reliability that is attractive in a world rife with uncertainty and ambiguity. However, this very feature of blockchain – its immutability – also raises critical challenges when interfaced with the mutable, subjective nature of human trust and interaction.

Human trust is a dynamic, context-dependent phenomenon. It is built, maintained, and sometimes broken and rebuilt over time, through a complex web of interactions, experiences, and emotions. This fluidity and adaptability of human trust present a stark contrast to the static certainty of blockchain. For instance, in the implementation of smart contracts within blockchain systems, the rigid, unchangeable nature of the contract terms can sometimes clash with the evolving, context-specific needs and understandings of the parties involved. Such scenarios reveal the limitations of a purely technological solution in addressing the nuanced requirements of human-centric trust.

The challenge, therefore, lies in finding a balance – in

integrating the security and reliability of blockchain with the flexibility and context-awareness that characterize human relationships. This involves not just a technological solution, but a rethinking of how we conceptualize trust in the digital age. It calls for a design philosophy that views technology not as a replacement for human trust, but as a complementary tool that enhances and supports the complexities of human interaction.

Delving into the realms of philosophy and sociology offers valuable insights into this challenge. Philosophers like Jürgen Habermas and sociologists like Anthony Giddens have long grappled with the relationship between technology and society. Habermas's discourse ethics, which emphasize the importance of communication and mutual understanding in the formation of societal structures, highlight the need for transparency, communication, and consensus in the integration of blockchain within society. Giddens's theories on modernity and self-identity echo the importance of considering the individual's role and identity in the face of advancing technology.

In the context of blockchain, these philosophical and sociological perspectives underscore the importance of designing systems that not only provide technical efficiency but also foster a sense of community, shared understanding, and mutual respect among users. This might involve mechanisms within blockchain platforms that allow for dialogue, dispute resolution, and adaptability in response to changing contexts and needs.

The paradox of blockchain in the realm of human trust is not a dilemma to be solved, but a dynamic interplay to be navigated. It presents an opportunity to rethink our approach to technology, not as a panacea for all societal challenges, but as a tool that, when thoughtfully integrated with human values and needs, can enhance the fabric of our interconnected lives. The future of blockchain, therefore, lies not just in its technical capabilities, but in our ability to harmonize these capabilities with the ever-evolving landscape of human trust and interaction.

The Philosophical Underpinnings of Blockchain and Trust

As we venture deeper into the paradox of blockchain and human trust, it becomes essential to delve into the philosophical foundations that underpin our understanding of this relationship. This section explores how the immutable nature of blockchain intersects with philosophical concepts of trust, ethics, and human interaction, drawing from the thoughts of influential philosophers to contextualize this modern technological phenomenon.

The philosophical examination of blockchain and trust takes us to the realm of Immanuel Kant and his categorical imperative. Kant's moral philosophy, which emphasizes actions based on duty and universal maxims, presents an interesting parallel to the blockchain's principle of transparency and ethical conduct. Blockchain, in its unalterable truthfulness, embodies Kant's ideal of acting in such a way that the principles of one's actions could become a universal law. However, the rigidity of blockchain's 'code is law' approach must be balanced with Kantian ethics' emphasis on human dignity and autonomy, ensuring that technological advancements serve humanity's broader ethical goals.

Michel Foucault's ideas about power structures and knowledge systems offer another lens through which to view blockchain. Foucault's analysis of how knowledge and power are intertwined in societal structures can be applied to blockchain's decentralized nature, which has the potential to disrupt traditional power hierarchies. The blockchain empowers individuals by providing transparent access to information, redistributing power in ways that align with Foucault's vision of a more egalitarian society. However, Foucault's caution about the pervasive nature of power systems serves as a reminder that technological advancements, while disruptive, can also create new forms of control and surveillance.

The exploration of trust and blockchain also aligns with the existentialist perspectives of philosophers like Jean-Paul Sartre and Martin Heidegger. Sartre's focus on individual freedom and responsibility resonates with the user empowerment that blockchain technology offers. It allows individuals to engage in transactions and interactions with a level of autonomy previously unseen. However, this freedom comes with the responsibility of understanding and navigating the complex world of blockchain, echoing Sartre's belief in the weight of individual choices.

Heidegger's reflections on technology and its impact on human existence also provide valuable insights. Heidegger warned of technology's potential to alienate individuals from authentic experiences, a risk that exists with blockchain's digitization of interactions. His advocacy for a more reflective relationship with technology urges a cautious integration of blockchain, ensuring that it enhances, rather than diminishes, human connections and experiences.

In synthesizing these philosophical perspectives, it becomes evident that blockchain technology, while revolutionary, must be approached with a nuanced understanding of its impact on human trust, ethics, and society. The philosophical discourse encourages a mindful application of blockchain, one that respects and enhances human values while harnessing its potential for transparency, decentralization, and empowerment. As we continue to navigate the complexities of integrating blockchain into our societal fabric, the guidance of these philosophical underpinnings will be instrumental in crafting a future where technology and humanity coexist in harmonious balance.

AARON VICK

Sociological Perspectives on Blockchain and Human Interaction

In exploring the blockchain paradox, it's crucial to consider the sociological implications of this technology on human interactions and trust. This section examines blockchain through the lens of sociological theories, analyzing how it influences and is influenced by the structures and dynamics of human society.

The Sociological Landscape of Blockchain:
Sociologists like Émile Durkheim and Max Weber offer frameworks to understand the societal impact of blockchain technology. Durkheim's concept of social solidarity and collective conscience can be applied to the decentralized nature of blockchain. Blockchain, by design, fosters a form of collective agreement and cooperation, echoing Durkheim's vision of a society bound by shared norms and practices. However, this technology also challenges traditional forms of societal organization, potentially leading to new forms of social order that Durkheim's theories help us to navigate.

Max Weber's analysis of rationalization and bureaucracy in modern societies provides another angle to comprehend blockchain's societal role. Blockchain can be seen as the epitome of Weber's rational-legal authority, where rules and procedures govern interactions rather than traditional hierarchies or charismatic leadership. This transition to a more rational and systematic form of organization, as facilitated by blockchain, could lead to more efficient and transparent societal systems but also raises questions about the depersonalization and disenchantment of social interactions.

Durkheim's Social Solidarity in the Blockchain Context:

In the digital realm of blockchain, the sociological theories of Émile Durkheim, particularly his concept of social solidarity, offer a compelling lens through which to view the emerging blockchain communities. These communities represent microcosms of society, each with distinct norms, values, and practices, united not just by the technology but also by a shared belief in principles like decentralization, transparency, and collective governance.

Durkheim's idea of organic solidarity, where different elements of society work in unison like organs in a body, each contributing to the overall health and stability, is strikingly evident in the way blockchain communities operate. Members from diverse backgrounds and skill sets come together, driven by a common purpose to support and advance the blockchain ecosystem. This unity isn't imposed through traditional hierarchies but arises naturally from shared interests and mutual dependencies.

Blockchain technology itself reinforces this organic solidarity. Its decentralized nature cultivates a sense of collective responsibility and active participation among community members. Each individual plays a crucial role in maintaining the network's integrity, echoing Durkheim's vision of a cohesive society built on specialized roles and contributions.

However, Durkheim also cautioned against the potential for anomie, a state of normlessness, especially in societies undergoing rapid change. This concern is particularly relevant in the context of blockchain. The technology's swift evolution and adoption can lead to uncertainties within communities, as traditional social norms and regulatory frameworks scramble to keep up. The lack of clear standards and governance models in the fast-evolving blockchain space can create disorientation and confusion among community members about appropriate behaviors and practices.

As blockchain permeates different sectors of society, the collective conscience of its communities – the shared beliefs and values that bind them together – may evolve or even clash with broader societal values. This poses a challenge for blockchain leaders and innovators to foster a collective conscience that not only aligns with the technology's principles but also resonates with wider societal values. The goal is to promote ethical practices, inclusivity, and a culture of cooperation and mutual respect, ensuring that as blockchain influences various aspects of society, it nurtures a healthy collective conscience within its communities, thereby maintaining social solidarity.

In sum, Durkheim's insights into social solidarity provide

a valuable framework for understanding and navigating the societal dynamics of blockchain communities. As we grapple with the challenges and opportunities posed by this transformative technology, Durkheim's theories highlight the importance of fostering organic solidarity, averting anomie, and cultivating a robust collective conscience in our digital age.

Weber's Rationalization and Blockchain Systems:

Max Weber's theories on rationalization and bureaucracy shed light on the adoption and integration of blockchain into societal structures. Blockchain, embodying Weber's concept of rational-legal authority, marks a significant shift from traditional authority forms. It emphasizes systems and procedures, heralding a more systematized and rational form of organization that enhances efficiency and fairness in societal processes.

In the realm of blockchain, Weber's rational-legal authority is vividly reflected. The technology is built on codified rules, consensus mechanisms, and immutable algorithms, aligning with Weber's vision of a society governed not by hierarchical or charismatic power but by rationalized processes. Authority in blockchain networks is derived from the systems and protocols themselves rather than from individuals or institutions. This represents a profound move towards a rational-legal form of governance, where decisions are made based on transparent rules and verifiable data.

Weber's analysis of societal rationalization highlights the potential of blockchain to streamline processes, promising efficiency and transparency beyond the capabilities of traditional systems. The technology's ability to automate and transparently record transactions addresses bureaucratic inefficiencies and corruption risks, resonating with Weber's idea of a rational, predictable, and efficient society.

However, Weber's critique of the 'iron cage' of rationalization presents a cautionary perspective for blockchain. He warned that excessive rationalization could lead to disenchantment, stripping social life of its meaning. Applied to blockchain, this raises concerns about the potential depersonalization of social interactions, reducing them to mere transactions within an automated system. The challenge lies in ensuring that blockchain's efficiency doesn't eclipse the need for human connection, meaning, and engagement in social interactions.

The solution involves a nuanced approach to blockchain implementation. While embracing its efficiency and reliability, it's crucial to design systems that consider social and emotional dimensions. Integrating human oversight, ethical frameworks, and community participation can help balance the rational efficiency of blockchain with the human aspects of social interactions.

Weber's theories offer a valuable framework for understanding blockchain's societal impact. While blockchain heralds a new era of rationalization and efficiency, it's imperative
to navigate its potential pitfalls. The goal is to develop blockchain systems that not only optimize efficiency but also
maintain a connection to the human experience, ensuring
that they support rather than undermine meaningful human
connections and interactions.

As we explore the intersection of blockchain technology
and sociological paradigms, the insights of Émile Durkheim
and Max Weber provide crucial frameworks for understanding its impact on social cohesion and organizational structure.

Émile Durkheim: Blockchain and Social Cohesion
Durkheim's concept of organic solidarity resonates profoundly within blockchain communities. These digital ecosys-
tems, characterized by decentralized and varied contributions,
mirror Durkheim's vision of a society where different roles
amalgamate for communal benefit. In blockchain networks,
each participant plays a unique role, yet all contribute to the
network's overall health and functionality. This decentralized
cooperation reflects a modern manifestation of Durkheim's
organic solidarity, where interdependence and specialized
roles foster collective well-being.

However, Durkheim's apprehensions about anomie – the
loss of social norms due to rapid societal changes – are

particularly relevant to the blockchain context. The swift pace of blockchain technology's evolution might outpace the development of new social norms and regulatory frameworks, leading to ethical ambiguities and a potential disintegration of traditional social structures. To mitigate this, blockchain communities must strive to establish and uphold a collective conscience, ensuring innovation is matched with a dedication to shared ethical principles and values.

Max Weber: Rationalization in the Blockchain Era

Weber's examination of rationalization and bureaucracy offers a complementary perspective on blockchain. The technology epitomizes Weber's notion of rational-legal authority, where systematic protocols and algorithms replace traditional hierarchical structures. This shift towards a more systematized and predictable form of organization suggests increased efficiency and fairness, aligning with Weber's concept of a rationalized society.

Yet, the challenge presented by Weber's 'iron cage' of rationalization – the potential loss of meaning and disenchantment in a highly rationalized world – is critical in the blockchain context. As blockchain systems promote efficiency and transparency, there's a risk that they might also depersonalize social interactions, reducing them to mechanized transactions. The challenge lies in balancing the streamlined efficiency of blockchain with the intrinsic human need for meaningful social connections.

Looking Ahead: Societal Implications of Blockchain

The application of Durkheim and Weber's theories to blockchain raises significant questions about the future of our social structures and interactions. As blockchain technology continues to evolve and integrate into various societal aspects, it's imperative to reflect on how it will reshape our social fabric – influencing social cohesion, ethical practices, and our sense of community and belonging.

The task ahead involves integrating blockchain technology into our societal framework in a manner that respects both its efficiency and the essential human values it should serve. This requires developing frameworks that support technological innovation while nurturing empathy, ethics, and shared societal goals.

In summary, the integration of blockchain into society represents not only a technological advancement but also a profound sociological and philosophical challenge. The way we embed this technology into our lives will significantly shape our social structures, collective values, and individual and communal identities in the digital age. The future of blockchain in society hinges on a balanced approach that honors both its potential for efficiency and the enduring importance of human connection and meaning.

Chapter 5: Work and Purpose in the Age of Automation and Blockchain

---

"The future of work is not about dull routine... it's about being more human." — Greg Satell

---

The landscape of work is undergoing a seismic shift with the advent of automation and blockchain technology. This transformation redefines not only job roles and employment patterns but also challenges the very essence of work as a central pillar of identity and societal structure.

Automation and AI have begun to take over tasks that were historically the domain of human labor, dramatically altering the job market and our conception of work. This evolution goes beyond economic implications, touching the core of personal identity and societal roles.

Blockchain technology adds a new dimension to this changing landscape. Its decentralized approach fosters more flexible, transparent work structures, from smart contracts to decentralized autonomous organizations (DAOs), offering a glimpse into a more equitable and participatory work environment.

This redefinition of work raises profound questions at both societal and individual levels. It brings forth opportunities for enhanced efficiency and creativity, yet also poses challenges regarding job security, income inequality, and the preservation of work-related social structures.

As we progress, the integration of AI and blockchain heralds a shift in job security and economic structures. Traditional notions of long-term employment are giving way to more dynamic, skill-based, and adaptable models of work. While this transition opens up new avenues for employment, it also necessitates a rethinking of economic stability and workforce support systems.

The chapter further explores how these technologies are reshaping societal norms and trust dynamics in the workplace, redefining principles of fairness, contribution, and trust in an increasingly digital world.

Transforming the Landscape of Work

The advent of automation and blockchain technology is profoundly reshaping the landscape of work. This transformation extends beyond mere changes in job roles or employment patterns; it signals a fundamental shift in the nature and meaning of work itself.

Automation and AI: Redefining Work Roles

Automation and artificial intelligence (AI) are rapidly automating tasks that were once considered the exclusive domain of human labor. From manufacturing assembly lines to customer service interactions, machines and algorithms are demonstrating remarkable capabilities in performing complex tasks with unmatched efficiency and precision. While this technological revolution undoubtedly promises to enhance productivity and economic growth, it also presents a significant challenge to our traditional understanding of work as a central pillar of identity and purpose.

The rise of automation and AI is not merely displacing workers from their traditional jobs; it is fundamentally altering the skills and competencies required for success in the modern workplace. The once-prized skills of manual dexterity and rote memorization are gradually giving way to a demand for creativity, adaptability, and problem-solving abilities. This shift necessitates a comprehensive reskilling and upskilling effort, empowering individuals to navigate the evolving job market and embrace the opportunities presented by this technological transformation.

Blockchain technology, on the other hand, is revolutionizing the way we organize work and manage value exchange. Its decentralized and transparent nature holds immense potential to address the challenges of trust, traceability, and efficiency that plague traditional business models. Blockchain-powered platforms are enabling new forms of collaboration, disrupting industries ranging from finance to supply chain management. This transformation is fostering a more equitable and inclusive work environment, empowering individuals to take ownership of their digital identities and participate in the global economy without intermediaries.

As we navigate this era of technological revolution, it is crucial to strike a delicate balance between the transformative power of automation and blockchain and the preservation of our core human values. We must ensure that our digital future is not merely technologically enriched but also profoundly human, guided by the principles of empathy, compassion, and mutual respect.

In this evolving landscape, the role of work is not diminishing but rather evolving. It is shifting from a mere means of subsistence to a platform for self-actualization, personal growth, and societal contribution. Technology is not replacing human labor; it is augmenting it, providing us with new tools and opportunities to redefine the ways in which we create value, connect with others, and contribute to the world around us.

The future of work is not one of technological dominance but rather of human empowerment. It is a future where technology and humanity coexist in harmony, not as opposing forces but as mutually reinforcing partners in shaping a world of limitless possibilities. In this future, work will not be about fulfilling predetermined tasks but about unleashing our creative potential, collaborating with others, and making a meaningful impact on the world. It will be a future where work is not just a job but a calling, a source of fulfillment and purpose.

Blockchain's Role in the New Work Paradigm

The advent of blockchain technology is fundamentally altering the nature of work, introducing new models of collaboration and organization that challenge conventional hierarchical structures. Its decentralized and transparent nature is enabling the emergence of more flexible and equitable work environments, empowering individuals to take ownership of their digital identities and participate in the global economy without intermediaries.

Smart Contracts and Decentralized Autonomous Organizations (DAOs)

Blockchain-powered smart contracts are self-executing agreements that automate the exchange of value and enforce predefined terms of engagement. This technology is revolutionizing the way freelance work is managed, ensuring transparency, fair compensation, and secure transactions between freelancers and clients.

Decentralized autonomous organizations (DAOs) are blockchain-based entities that operate without centralized leadership, relying on consensus-driven decision-making processes. DAOs are pioneering new models of corporate governance, enabling individuals to participate directly in the decision-making and operations of organizations.

The Rise of Gig Economies and Decentralized Work Models

The convergence of AI and blockchain technologies is driving the growth of gig economies, characterized by flexible, project-based work arrangements. These economies are providing individuals with greater autonomy and opportunities to monetize their skills, while also challenging traditional notions of employment and job security.

Blockchain's inherent transparency and security are fostering trust in these decentralized work environments. By providing an immutable record of transactions and ensuring fair compensation, blockchain is empowering individuals to engage in gig work with greater confidence and protection.

Blockchain's Transformative Impact on the Gig Economy

The gig economy, characterized by flexible, project-based

work arrangements, is undergoing a profound transformation driven by the convergence of AI and blockchain technologies. These technologies are introducing new models of collaboration, organization, and payment, empowering individuals to take ownership of their digital identities and participate in the global economy without intermediaries.

Transparency and Security in Freelance Work: The Case of Upwork

Upwork, a leading online marketplace for freelance services, has integrated blockchain technology to enhance its platform. By utilizing blockchain for payments and escrow, Upwork aims to provide greater transparency and security for both freelancers and clients. This integration addresses a longstanding challenge in the gig economy – the lack of trust between freelancers and clients regarding payments and project completion. Blockchain's immutable ledger ensures an indisputable record of transactions, eliminating the risk of fraud or disputes. Additionally, smart contracts automate the payment process, ensuring that freelancers are paid promptly and fairly upon completion of their work.

Automating Hiring and Payment for Temporary Workers: The LaborX Paradigm

LaborX, a decentralized marketplace for temporary labor, leverages blockchain technology to automate the hiring process and ensure fair compensation for workers. The platform uses smart contracts to match employers with suitable workers, eliminate intermediaries, and automatically execute payments based on pre-agreed terms. This automation streamlines the hiring process for businesses, reducing administrative burdens and ensuring timely payments for workers. Furthermore, LaborX's smart contracts provide transparency by ensuring that both employers and workers have access to a verifiable record of all transactions.

The Transformative Impact of Blockchain on the Gig Economy

Blockchain technology is revolutionizing the gig economy by introducing transparency, security, and automation. These advancements are addressing long-standing challenges in the decentralized work model, fostering trust between participants and enhancing the efficiency of transactions. As blockchain applications continue to develop, we can expect to see even more innovative ways in which this technology will shape the future of work.

Key Takeaways

- Blockchain technology is introducing new models of collaboration, organization, and payment in the gig economy.
- Upwork's integration of blockchain enhances transparency and security in freelance work.
- LaborX's blockchain-based platform automates hiring and ensures fair compensation for temporary workers.
- Blockchain's transformative impact on the gig economy is fostering trust, efficiency, and innovation.

Addressing the Challenges of a Changing Work Landscape

The shift towards gig economies and decentralized work models raises concerns about long-term financial security, access to benefits, and the potential for increased income inequality. Addressing these challenges requires innovative thinking and policy interventions that can adapt to and support these new economic realities.

Policies that promote lifelong learning, provide support for reskilling and upskilling, and establish universal basic income or social safety nets can help mitigate the negative impacts of a changing job market. Additionally, regulatory frameworks that ensure fair treatment and protection for gig workers are essential to fostering a more equitable and inclusive work environment.

Redefining Fairness, Contribution, and Trust in the Digital Age

As AI and blockchain technologies continue to reshape the world of work, it becomes increasingly important to re-examine the principles of fairness, contribution, and trust. In a decentralized and increasingly digital world, these principles need to be adapted to ensure that the benefits of technological advancements are distributed equitably and that individuals are empowered to participate meaningfully in the global economy.

Fairness must encompass not only equal access to

opportunities but also fair compensation, protection from exploitation, and the ability to challenge unfair practices. Contribution must extend beyond traditional employment models, recognizing and valuing the diverse ways in which individuals contribute to society, whether through paid work, unpaid caregiving, or creative pursuits.

Trust, in this evolving landscape, must be built on transparency, accountability, and the ability to hold both individuals and organizations responsible for their actions. Blockchain technology, with its inherent transparency and immutability, can play a crucial role in establishing trust in decentralized work environments.

By embracing these principles and harnessing the transformative power of AI and blockchain, we can create a future of work that is not only more efficient and flexible but also more equitable, inclusive, and empowering, ensuring that the benefits of technological advancements are shared by all.

Concrete Example: DAOs in Action

One of the most compelling illustrations of the transformative impact of blockchain technology on work paradigms is the rise of Decentralized Autonomous Organizations (DAOs). Platforms such as Aragon or DAOstack exemplify how blockchain can revolutionize organizational structures and governance models.

In these DAOs, blockchain technology enables a radical

departure from traditional hierarchical management, fostering a more democratic and equitable work environment.

In DAOs, individuals from various backgrounds and locations come together to collaborate on projects, make collective decisions, and share in the profits. These organizations operate on principles of transparency and equality, with each member having a stake and a voice in decision-making processes. This structure is in stark contrast to conventional top-down organizational models and introduces a new way of working and collaborating.

Take, for instance software development, DAOs offer a collaborative space where developers can work on open-source projects. Unlike traditional work environments, these developers are not bound by company policies or limited by geographical constraints. They contribute to projects that resonate with their interests and skills, fostering a sense of purpose and community. The transparent and equitable distribution of rewards, managed through blockchain, ensures that each contributor's effort is recognized and compensated fairly.

Example DAOs in action demonstrate the potential of blockchain to create more inclusive and participatory economic structures. They are not just platforms for employment but are also communities where individuals can engage, learn, and grow together. The success of DAOs in various fields highlights the potential of blockchain to facilitate collaborative and decentralized work environments, challenging traditional notions of organizational structure and employment.

As we delve further into the societal norms and trust dynamics in these changing work paradigms, it becomes clear that the principles of fairness, contribution, and trust are being redefined in an increasingly blockchain-influenced world. DAOs exemplify how blockchain can be leveraged to create not only more efficient and transparent work environments but also more democratic and socially responsive economic systems.

Addressing Societal Norms and Trust

The integration of blockchain and artificial intelligence (AI) into the workforce marks a significant shift in societal norms and the nature of trust in professional environments. Blockchain technology, known for its transparency and immutability, introduces a new level of trust in digital transactions and collaborations. This shift is gradually redefining the traditional workplace dynamics and the way trust is built and sustained in professional settings.

Blockchain's capacity to provide a transparent and immutable record of transactions and interactions offers a form of trust that is rooted in verifiability and security. This form of trust is particularly significant in environments where transparency and accountability are paramount, such as in financial transactions, supply chain management, or data handling. In such scenarios, blockchain acts as a neutral and reliable third party, ensuring that the records of transactions and interactions are tamper-proof and easily verifiable by all parties involved.

However, this technological advancement also brings forth challenges in maintaining personal trust and loyalty, which have been central to traditional workplace relationships. The impersonal nature of blockchain and AI systems could potentially lead to a depersonalization of work interactions. Where trust was once built through face-to-face interactions, shared experiences, and a mutual understanding of individual strengths and weaknesses, it now risks being reduced to reliance on algorithmic processes and digital records.

This shift calls for a nuanced approach to integrating blockchain into the workplace. It's essential to balance the efficiency and objectivity provided by these technologies with the human aspects of trust, which are built on empathy, understanding, and interpersonal relationships. For instance, while blockchain can streamline processes and enhance transparency in a project management system, it's crucial to ensure that the team members still engage in meaningful interactions and collaborations, fostering a sense of community and mutual trust.

The role of AI and blockchain in reshaping workplace norms extends beyond the technology itself to encompass the broader implications on social structures and values. As these technologies automate various aspects of work, they also challenge the traditional norms of job roles, career paths, and workplace hierarchies. Adapting to these changes requires not only a technological adjustment but also a cultural and societal shift in how we perceive work, value contributions, and build trust within and across organizations.

Reimagining Trust in the Blockchain Era

The evolution unfolding in the realms of blockchain and artificial intelligence is a complex interplay of technology, philosophy, and society, profoundly challenging and reshaping our long-established beliefs about the foundations of trust. It marks a departure from the conventional understanding, steering us into an era where trust is not just a human intuition but also a product of algorithmic certainty and cryptographic security.

In this era, the definition of trust expands beyond the emotional and intuitive, becoming intertwined with the precise and the predictable. The philosophical implications are significant, as we must now grapple with how these technological advancements either align or conflict with traditional notions of trust, historically rooted in human interactions and personal experiences.

Achieving a delicate balance in this new paradigm is critical. On one side of the scale lies the efficiency and certainty offered by technology, and on the other, the inherently human qualities of empathy, understanding, and emotional connection. This balance is vital for ensuring that technological advancements enhance, rather than diminish, the richness of the human experience.

The journey into this new digital landscape is fraught with ethical challenges. Key among these is ensuring that the systems we build and adopt are not only efficient and secure but also equitable, just, and respectful of human dignity. The development and application of blockchain and AI must be underpinned by our collective societal values and aspirations, guiding these technologies towards positive and ethical use.

Our vision for the future is one where technology and humanity coexist in a symbiotic relationship. In this envisioned future, blockchain and AI are not mere tools for economic and operational efficiency but also instruments for reimagining our social structures, relationships, and the essence of trust. This is not solely a technological journey but a deeply human one, where shared ethics, values, and aspirations shape a future that elevates the human experience in the age of automation and blockchain.

In this landscape, blockchain and AI become more than technological marvels; they are catalysts for a new era of human interaction and understanding. It's a future where digital advancements do not undermine our sense of community and connection but enrich it, creating a world that is more interconnected, transparent, and grounded in trust. As we navigate this evolving terrain, our collective goal is to harness these technologies in ways that celebrate and enhance our shared human journey, crafting a world where digital progress is inextricably linked with the advancement of human values and ethics.

Chapter 6: Redefining Societal Contracts in the Blockchain World

---

"Social contracts need to be rewritten... now that we have the ability to coordinate on a global scale." — Vitalik Buterin

---

Blockchain technology brings a profound shift, distributing trust across a network rather than centralizing it in a single entity. This decentralization not only changes how trust is established and maintained but also offers the potential for more democratic and equitable systems. The implications of this transition are vast, reshaping societal operations and creating new opportunities for governance and social structures.

A key focus of this chapter is the exploration of

Decentralized Autonomous Organizations (DAOs) and their role in redefining governance models. DAOs like Aragon and DAOstack are pioneering examples of how blockchain can facilitate collective decision-making, eliminating centralized control in favor of a democratic approach. This new model of governance, operating through blockchain's consensus mechanisms, challenges traditional hierarchical structures and proposes a future of governance that is more responsive to community needs.

However, the move towards blockchain-based societal contracts is not without its challenges. Critical concerns arise around the ethical implications of embedding human values into technological systems. The design and implementation of blockchain must transcend technological efficiencies to encompass principles of equity, justice, and inclusivity. This balance between blockchain's efficiency and the preservation of human-centered values is crucial in ensuring that the technology serves as a positive force for societal transformation.

In redefining societal contracts through blockchain, we are compelled to address not only the technological potential but also the ethical and practical challenges accompanying it. The goal is to create a societal framework that leverages the strengths of blockchain while upholding values fundamental to a just and equitable society.

AARON VICK

Revisiting Societal Contracts in the Age of Blockchain

Traditional societal contracts, once reliant on centralized authorities and institutions for trust and agreement enforcement, face a paradigm shift with the decentralized and transparent nature of blockchain. This technology introduces a radical departure from the norm, distributing trust across a network, thereby reshaping societal interactions from institutional reliance to faith in technology and protocol.

This seismic shift in societal dynamics unlocks opportunities for more democratic and equitable systems, challenging long-standing governance models and proposing new paradigms of collective ownership and decision-making. Decentralized Autonomous Organizations (DAOs), like Aragon and DAOstack, exemplify this transformation. Operating without centralized control, these organizations rely on collective decision-making through consensus mechanisms intrinsic to blockchain, suggesting a future of governance that is more attuned to community needs and values.

The implementation of these blockchain-based governance models across various sectors demonstrates a significant move towards transparent and equitable decision-making. Yet, this transformation is not without its ethical implications and challenges. The integration of human values into blockchain systems is critical, ensuring they embody not just technological efficiencies but also principles of equity, justice, and inclusivity. This raises the question of how to balance the benefits of blockchain, like transparency and efficiency, with the preservation of human-centered values.

In the broader context of societal governance, blockchain's potential is far-reaching. It can facilitate direct democracy models, enabling community members to vote on issues directly, enhancing democratic processes, and making governance more responsive to public needs. Estonia's e-Residency program and land registry systems in countries like Georgia and Sweden are practical examples of blockchain's application in public administration, showcasing its ability to create more inclusive and accessible governance models.

However, these advancements come with their own set of challenges and ethical considerations. Issues of accessibility, privacy, and security are paramount in the design and implementation of blockchain systems in governance. There is also a pressing need to ensure that these systems do not contribute to digital divides or enable privacy infringements and surveillance.

In summary, the role of blockchain in shaping new societal norms and governance models is profound and complex. It offers opportunities for more transparent, efficient, and inclusive governance structures, contributing significantly to the evolution of societal contracts. Navigating this transformation requires a thoughtful and balanced approach, addressing the ethical and practical challenges inherent in integrating this technology into society's fabric.

The future of blockchain in shaping societal norms and governance models is a journey of careful consideration, collaboration, and adaptation, ensuring that its integration serves the broader goals of societal welfare and individual freedoms.

Blockchain's Role in Shaping New Societal Norms and Governance Models

Blockchain technology is ushering in a new era in societal governance, transforming traditional models and introducing innovative structures powered by smart contracts. This shift heralds a future where governance is not only more transparent and efficient but also more participatory, fundamentally altering how decisions are made and executed.

A notable aspect of this transformation is the potential for direct democracy models, facilitated by blockchain's ability to securely and transparently record votes. This capability democratically empowers communities, allowing members to directly influence decision-making processes without intermediaries. Such an approach revitalizes democratic practices, making governance more responsive to the needs and opinions of the populace.

In the realm of public administration, blockchain's transformative impact is evident. Estonia's e-Residency program is a pioneering example, leveraging blockchain to provide secure digital identities. This initiative has set a precedent for how blockchain can create more inclusive and accessible governance models, transcending geographical boundaries and fostering global participation in the digital economy.

Similarly, blockchain's exploration in land registry systems in countries like Georgia and Sweden exemplifies its potential to enhance governance efficiency. These systems aim to increase transparency, reduce fraud, and streamline property transactions, thereby building public trust and improving administrative processes.

However, the integration of blockchain into governance is not without its challenges and ethical considerations. As blockchain systems are developed, it is critical to address issues of accessibility, privacy, and security. Ensuring that these systems are inclusive and do not contribute to widening digital divides is paramount. Additionally, there is a pressing need to carefully consider the potential of blockchain in ways that might infringe upon privacy or enable surveillance, balancing the benefits of efficiency and transparency with the protection of individual rights and freedoms.

In conclusion, blockchain's impact on societal norms and governance models is profound, offering the promise of more transparent, efficient, and inclusive structures. However, the path to realizing this potential is complex, requiring a careful and thoughtful approach. This involves addressing the ethical and practical challenges inherent in integrating this technology into society's fabric. As we navigate this transformation, the goal remains clear: to leverage blockchain in a way that enriches societal structures while aligning with the values and principles fundamental to a just and equitable society.

In the transformative realm of blockchain technology, its potential to redefine social justice, equity, and participation emerges as a cornerstone of its societal impact. The decentralized and transparent nature of blockchain creates new avenues for addressing corruption, inequality, and ensuring fair practices, marking a significant shift in how societal contracts are understood and enacted.

Empowering Social Justice and Transparency

Blockchain's ability to ensure transparent and equitable systems presents it as a powerful tool in the fight against societal injustices. Its application in charity and non-profit sectors exemplifies this, where it is used to track donations, guaranteeing that funds are utilized effectively and reach their intended recipients. This transparency builds trust with donors and enhances the effectiveness of social programs.

Fair Trade and Equality

In fair trade initiatives, blockchain is transformative, ensuring traceability in the supply chain. This traceability guarantees that producers, particularly in developing countries, receive fair compensation for their labor. Consumers are empowered to make informed decisions, supporting ethical practices and fair labor standards. This application of block-chain in ensuring fair trade and equality demonstrates its potential in promoting social equity.

Transforming Participation in Society

Blockchain technology also redefines societal participation, especially for marginalized or underrepresented groups. By providing secure digital identities, it facilitates greater access to essential services like voting, healthcare, and financial services. This aspect is particularly empowering for individuals who lack traditional forms of identification, broadening their participation in societal processes.

Ethical Considerations and Challenges

While blockchain presents numerous opportunities, it also raises important ethical considerations. The design and implementation of blockchain systems must prioritize inclusivity and guard against exacerbating digital divides. Potential misuses of blockchain in surveillance or control mechanisms pose significant ethical concerns. Ensuring that blockchain applications align with principles of social justice and equity is paramount.

Envisioning a Blockchain-Influenced Society

Looking ahead, the focus is on harnessing blockchain tech-nology for societal betterment. Thoughtful integration into societal systems presents the opportunity to create a more just, equitable, and participatory world. However, achieving this requires a collective effort from technologists, policymakers, and society at large. The potential of blockchain to redefine societal contracts lies in our collective vision and actions to utilize this technology ethically and equitably.

In envisioning a future society influenced by blockchain, the focus must be on harnessing the technology for the greater good. By thoughtfully integrating blockchain into societal systems, there is an opportunity to create a more just, equitable, and participatory world. However, this requires a collective effort from technologists, policymakers, and society at large to ensure that the benefits of blockchain are distributed equitably and ethically. The potential of blockchain to redefine societal contracts lies not just in its technological capabilities but in our collective vision and actions to use it for the betterment of society.

Chapter 7: Trust in a Decentralized World: Opportunities and Challenges

---

"Decentralization is not just a technological problem, it's a human problem." — Amir Taaki

---

Historically, societal trust has been anchored in centralized institutions like banks and governments, entities that have stood as pillars of reliability and authority. Blockchain technology challenges this centralization, dispersing trust across a network of users and thus diminishing the hegemony of traditional institutions. This paradigm shift ushers in a new era of transparency and diminishes opportunities for corruption and fraud, marking a significant evolution in how trust is established and maintained in society.

The decentralization intrinsic to blockchain technology brings forth a spectrum of opportunities, particularly in democratizing access to power and resources. The rise of decentralized finance (DeFi) platforms exemplifies this, reshaping access to financial services and negating the necessity of traditional intermediaries. This democratization extends beyond finance, influencing various facets of societal interaction and governance.

However, the journey towards a decentralized trust landscape is not without its challenges. Key among these are the security concerns inherent in decentralized networks, which are susceptible to a range of cyber threats, and the psychological shift required to place trust in algorithms and community consensus over familiar institutions.

Redefining Trust in Decentralized Systems

The emergence of blockchain technology signifies a seismic shift in the nature and structure of trust. This section delves into the intricate dynamics of trust in a decentralized world, examining the opportunities and challenges presented by this paradigm shift.

The Paradigm Shift in Trust Structures:

Blockchain technology's decentralized framework marks a significant departure from the conventional trust systems that have been traditionally centralized in established institutions like banks, corporations, and governments. In decentralized blockchain networks, trust is not reliant on a single central authority; instead, it is distributed across a network of participants. This reconfiguration of trust represents a fundamental shift from the historical reliance on centralized entities to guarantee and mediate transactions and interactions.

The decentralized nature of blockchain offers a new form of trust, one that is built into the very architecture of its technology. Transactions and interactions on a blockchain are verified and recorded on a public ledger, accessible to all participants, providing an unprecedented level of transparency. This transparency is crucial in reducing opportunities for corruption and fraud, as every action or transaction within the network is recorded and cannot be altered retrospectively.

Challenges in a Decentralized Trust Environment:
While the decentralization of trust offers many benefits, it also introduces several challenges. One of the primary challenges is the shift in the perception of trust. In traditional systems, trust is often associated with familiar and established institutions. In a decentralized environment, trust must be established in the technology and the algorithms that govern the blockchain. This shift requires a new understanding and acceptance of technology as a reliable mediator of transactions and interactions.

Another challenge in decentralized systems is ensuring the security and integrity of the network. While blockchain's design inherently offers robust security features, such as cryptography and consensus mechanisms, the system is not immune to vulnerabilities. The decentralized nature means that if a part of the network is compromised, it could potentially impact the entire system. Ensuring the ongoing security and resilience of decentralized networks is, therefore, a critical area of focus.

Decentralization: A New Paradigm in Societal Interaction

Blockchain introduces a radical shift from the traditional reliance on centralized authorities to a distributed network approach. This shift alters the dynamics of societal interactions, fostering a move from institutional trust to a reliance on technology and protocol. The implications of this transformation are far-reaching, offering a foundation for more democratic and equitable systems. For instance, in decentralized finance (DeFi), blockchain enables financial activities without traditional intermediaries, democratizing access to financial services and empowering previously unbanked or underbanked populations.

Redefining Governance with Decentralized Autonomous Organizations (DAOs)

The concept of Decentralized Autonomous Organizations (DAOs) encapsulates the transformative power of blockchain in governance. Operating without centralized control, DAOs rely on collective decision-making by members, challenging traditional hierarchical governance models. This new approach to governance, exemplified by organizations like Aragon or DAOstack, is more than theoretical; it's being implemented across sectors, fostering a shift towards transparent and equitable decision-making processes.

Ethical Implications and the Balancing Act

Despite its transformative potential, the integration of blockchain into societal structures raises significant ethical concerns. The challenge lies in embedding human values into these technological systems, ensuring they reflect principles of equity, justice, and inclusivity. A critical balance must be struck between leveraging blockchain's efficiencies and preserving human-centered values to prevent exacerbating inequalities or creating new forms of exclusion.

Future Outlook: Redefining Societal Contracts with Blockchain

As blockchain continues to integrate into various aspects of society, it compels a reevaluation and redefinition of our societal contracts. This process involves embracing blockchain's technological potential while addressing the ethical and practical challenges it presents. The goal is to establish a societal framework that capitalizes on blockchain's strengths while upholding the values and principles fundamental to a just and equitable society. This journey towards a blockchain-influenced society is not just about technological advancement but about shaping a future where technology harmoniously aligns with human values and societal needs.

As we navigate this transformative landscape, it becomes imperative to design blockchain systems that not only offer unparalleled security and transparency but also enhance and complement traditional trust mechanisms, ensuring a seamless integration of technology into society's trust structures.

At the core of this evolution lies the recognition that while blockchain can significantly bolster security and transparency, it is not a standalone solution to all trust-related societal challenges. True trust in human interactions – characterized by emotional depth and relational nuances – cannot be entirely replicated by any technology.

Therefore, blockchain systems must be seen as augmentative, enhancing existing trust structures rather than replacing them. This approach calls for human-centric design in blockchain systems, where the focus extends beyond mere technical efficiency to the impact on individuals' lives and relationships. By creating user experiences that are intuitive and accessible, and engaging diverse communities in the development process, these systems can become more inclusive and resonant with a broader spectrum of societal needs.

Ethical considerations are pivotal in the development of decentralized systems. Ensuring user privacy, promoting equity, and avoiding the exacerbation of social divides are essential ethical tenets. Moreover, the governance of blockchain net-works should embody democratic values, promoting inclusive decision-making and fair representation. Educational initiatives play a crucial role in easing the cultural and psychological shift towards decentralized trust. Raising awareness about blockchain's potential and limitations, and fostering community engagement, can help bridge the knowledge gap, encouraging widespread understanding and adoption.

The future of decentralized systems will be shaped by the ongoing dialogue between technological advancements and human values. This future will be one of continuous experimentation, adaptation, and vigilance, ensuring that as we embrace blockchain's opportunities, we remain mindful of its challenges and impacts. The overarching goal is to create a decentralized trust landscape that leverages blockchain's strengths while enhancing the human dimensions of trust, community, and shared values.

In summary, the journey towards a balanced approach in decentralized systems is an evolving narrative. It requires a collective effort from all societal sectors to ensure that the benefits of decentralization are realized in a manner that respects and enriches our human experience. The integration of blockchain into our trust structures is not just a technological leap but a step towards a future where technology empowers and amplifies the essence of our shared human journey.

Decentralization and its Societal Implications

In the context of a decentralized world, blockchain technology not only redefines trust but also reshapes societal structures and interactions. This section examines the broader societal implications of decentralization, focusing on its impact on governance, social norms, and individual agency.

Decentralization Transforming Governance Models:
Decentralization, as facilitated by blockchain, introduces new models of governance that challenge traditional hierarchical structures. Decentralized Autonomous Organizations (DAOs) exemplify this shift. In DAOs, governance is exercised through collective decision-making and consensus mechanisms embedded in blockchain technology. This model fosters a more egalitarian and participatory form of governance, empowering communities and individuals to have a more direct role in decision-making processes.

Blockchain in Municipal Governance - The Barcelona Example:
Barcelona's foray into blockchain technology for municipal governance represents a pioneering effort in reimagining how city administration can function in the digital age. The city's initiative to integrate blockchain into various public services aims to foster a more dynamic, responsive, and citizen-centric governance model. This approach is rooted in the principle of making city administration more accountable and accessible to its residents.

Barcelona's innovative journey in integrating blockchain technology into its municipal governance is a testament to how cities can harness digital advancements to enhance citizen engagement, streamline administrative processes, and ensure transparent public service delivery. This forward-thinking approach has positioned Barcelona as a leader in applying blockchain for societal governance.

The city's use of blockchain has revolutionized citizen engagement. By incorporating this technology into public voting, community decision-making, and feedback collection, Barcelona's residents can directly interact with city officials. This secure and transparent communication channel empowers citizens to voice their concerns and actively contribute to policy-making, fostering a more participatory and inclusive governance model.

On the administrative front, blockchain has significantly streamlined processes such as permit applications, document verification, and public record maintenance. The impact is profound – reduced waiting times, lower administrative costs, and a decrease in bureaucratic red tape. This shift not only enhances the efficiency of municipal operations but also elevates citizen satisfaction, contributing to a smoother, more efficient civic experience.

Transparency in public service delivery is another hallmark of Barcelona's blockchain implementation. The city leverages the technology's inherent transparency to meticulously track and report the progress of public projects and the use of municipal resources. This level of openness nurtures trust among residents, who gain a clearer understanding of how their taxes and public funds are being utilized for the city's development.

A critical aspect of Barcelona's blockchain system is its focus on secure data management. The city places a high priority on the confidentiality and integrity of citizens' personal information. Blockchain's secure framework ensures that sensitive data is protected from unauthorized access and breaches, safeguarding residents' privacy.

The broader implications of Barcelona's blockchain implementation extend beyond its city limits, signaling a shift in how local governments worldwide can leverage technology for better governance. By decentralizing decision-making and enhancing transparency, blockchain empowers citizens and builds trust in public institutions, paving the way for more responsive and accountable governance.

However, the journey is not without its challenges. A primary concern lies in ensuring that all citizens, regardless of their digital literacy or socio-economic status, have equitable access to blockchain-based services. Additionally, the city faces the complex task of integrating this emerging technology with existing legacy systems and aligning it with regulatory standards.

The case of Barcelona's blockchain initiative in municipal governance serves as a model for other cities and governments exploring similar transformations. It demonstrates how blockchain technology can be leveraged not just for economic transactions, but as a tool for enhancing democratic governance, citizen participation, and public service efficiency. As blockchain continues to evolve, its potential to reshape the landscape of municipal governance and public administration globally becomes increasingly apparent.

Impact on Social Norms and Community Dynamics:

The shift towards decentralized systems also impacts social norms and community dynamics. By decentralizing trust and authority, blockchain enables more fluid and dynamic social interactions. It encourages the formation of new types of communities united by shared interests and values rather than geographical proximity or institutional affiliations. This evolution in social norms reflects a move towards a more networked and collaborative society, where traditional power dynamics and social hierarchies are challenged.

Blockchain's role extends to revolutionizing decision-making processes, transitioning them from traditional hierarchical structures to more democratic and participatory models. Structures like Decentralized Autonomous Organizations

(DAOs) exemplify this shift, where collective decision-making is facilitated through blockchain's consensus mechanisms rather than centralized control. This evolution in decision-making reflects a broader societal transformation towards more democratic and equitable systems.

The potential of blockchain to democratize financial services is vividly illustrated in the realm of decentralized finance (DeFi). These platforms enable individuals to engage in financial activities traditionally mediated by banks or financial intermediaries, thus democratizing access to financial services and fostering financial inclusion for unbanked or underbanked populations. This shift is not just about efficiency but also about equity, allowing for more equitable participation in economic and societal systems.

Despite the numerous opportunities, the transition to decentralized trust systems presents substantial challenges. Ensuring robust security in these systems, where control and decision-making are distributed, is paramount. The move from institution-based trust to algorithmic and community-based trust necessitates considerable cultural and psychological adjustment. This requires a comprehensive understanding of blockchain technology and a reevaluation of longstanding beliefs about trust and security.

The future of decentralized trust systems hinges on successfully addressing these challenges. Educating users about the workings of decentralized systems and the nature of the risks involved is crucial. Developing frameworks and tools that provide users with a sense of security and recourse is also essential. By fostering a secure, transparent, and user-friendly environment, decentralized systems have the potential to transform the landscape of trust, offering more equitable and democratic alternatives to traditional centralized systems.

Chapter 8: Future Visions: Humanity and Technology in Harmony

"In the end, it's not about the technology; it's about the story it helps you tell." — Shivvy Jervis

Envisioning a future where blockchain technology seamlessly melds with the fabric of human trust and values, this chapter embarks on a journey to explore a world where digital advancements and humanistic principles exist in perfect synergy. Here, we delve into how the immutable nature of blockchain can complement the evolving dynamics of human society, aiming to create a future where technology serves as a cornerstone for enhanced trust, deeper connections, and societal progress.

In this envisioned world, the principles of blockchain – decentralization, transparency, and security – interweave with the very essence of human interaction, empathy, and ethical responsibility. We explore scenarios where blockchain goes beyond its current financial and transactional applications, becoming an integral part of diverse sectors such as healthcare, education, and governance.

This chapter examines the potential for blockchain to not only revolutionize systems and processes but also to uphold and strengthen the social and moral fabric of society. We imagine a world where technological progress and human well-being are not just parallel goals but are deeply inter-connected, each driving the other towards a more equitable and connected global community.

Envisioning a harmonious integration of blockchain tech-nology with human trust and values presents a captivating future. This fusion aims to uplift human experience, align-ing technological innovation with humanity's broader aspira-tions. We imagine blockchain not just as a digital tool but as a catalyst for profound human connections, where its princi-ples of transparency, decentralization, and security harmonize with empathy, community spirit, and ethical responsibility.

This future seeks to balance the precision of block-chain systems with an acute sensitivity to human needs and emotions, emphasizing technology that mirrors and respects human values. It's about molding blockchain into an intrinsic part of our daily lives, enriching interactions subtly yet significantly, without overshadowing the human element.

Such a vision calls for collective effort, bridging the gap between technologists, policy makers, sociologists, and the community. It involves ongoing dialogue to shape technology in a way that resonates with human needs and aspirations, ensuring technological progress is integrated into the societal fabric. Central to this vision is empowering individuals to interact confidently with blockchain technology, not just through technical expertise, but with an understanding of its societal implications and ethical dimensions.

Blockchain's role in this envisioned future is pivotal, serving as a foundation for secure, transparent, and equitable interactions in various life aspects, from governance to commerce to social engagement. Its influence goes beyond technical utility, contributing to a globally connected, empathetic, and understanding society. This future is not merely about leveraging technology for efficiency, but about using it as a bridge to a more interconnected and empathetic world.

Ethical Considerations and Philosophical Musings on a Tech-Augmented Future

As we journey through the realm of blockchain and its integration into the fabric of society, the indispensable role of interdisciplinary collaboration becomes strikingly clear. This collaborative spirit, drawing upon the expertise and perspectives of technologists, ethicists, policymakers, and end-users, is fundamental in ensuring that the progression of blockchain technology aligns with ethical standards and societal needs.

At the heart of this collaborative approach lies the synergy between diverse fields. Technologists contribute their understanding of what's feasible, ethicists weigh in on moral implications, policymakers shape legal frameworks, and end-users provide insights on practical impacts. This melding of different perspectives nurtures a shared vision for technology's role in society that extends beyond technical achievements to encompass ethical and societal considerations.

In designing technology, particularly in the blockchain domain, this interdisciplinary nexus proves crucial. Collaboration between technologists and ethicists ensures that human dignity and privacy are respected, guiding the creation of responsible and ethical technological solutions. Similarly, the dialogue between policymakers and technologists is essential for developing dynamic legal frameworks that foster innovation while protecting societal welfare.

Engaging end-users in the development process and initiating comprehensive educational initiatives are vital steps towards building trust and acceptance of new technologies. It is through this engagement that technological solutions can be aligned with societal values and user needs, ensuring the development of a more connected, transparent, and equitable world.

As we forge ahead, we are confronted with unprecedented ethical dilemmas brought forth by the rise of blockchain technology. Our approach is characterized by guiding innovation with ethical principles, ensuring that each technological advancement serves the greater good without compromising individual rights and societal values. We strive to develop solutions that are empathetically attuned to human interactions and societal shifts, reflecting a development ethos that is both technically proficient and socially conscious.

Creating adaptive governance models and learning from historical contexts are essential for managing the ethical ramifications of blockchain technology. As we move forward, engaging diverse voices and building strategic alliances across different sectors will be key in navigating the complex interplay between technology and societal norms.

In conclusion, our journey towards a future where technology and humanity coexist in harmony is an ongoing endeavor. It requires a commitment to ethical technology development, thoughtful regulation, and education. By embracing a balanced approach, we aim to ensure that technological advancements enrich the human experience and uphold our collective values. This approach will enable us to create a future where technology amplifies, rather than diminishes, the essence of our humanity.

Chapter 9: The Path Forward

"The good life is a process, not a state of being. It is a direction, not a destination." — Carl Rogers

---

In the realm of technological advancements, blockchain technology emerges as a beacon of hope, illuminating a path towards a future where the digital and human realms intertwine in profound harmony. This isn't merely about technological evolution; it's a profound shift in how we perceive and interact with the digital world, a transformation that has the potential to reshape not just our societies but also our very understanding of what it means to be human.

As we stand on the precipice of possibilities like consciousness replication or digital existence, blockchain emerges as a crucial tool, not just for its original financial applications but as a framework for safeguarding our humanity in an increasingly digital landscape. This exploration transcends the conventional applications of blockchain, venturing into a quest to understand how this technology can be instrumental in shaping a digital future where our human essence is not just mirrored but amplified.

The real challenge lies in harnessing blockchain's transformative potential to foster a future where digital advancements and human consciousness coexist, enriching rather than eclipsing our human experience. As we venture into this uncharted digital future, the focus shifts to the ethical, philosophical, and humanistic implications of such advancements. Blockchain, in its intricate design and inherent possibilities, offers a unique opportunity to anchor our humanity in the digital realm.

Blockchain's decentralized and transparent nature holds the promise of establishing a trustworthy digital infrastructure, one that can be built upon without compromising the fundamental values that define our humanity. This technology can empower individuals to take control of their digital identities and data, fostering a sense of ownership and self-determination in the digital world.

Furthermore, blockchain's potential to facilitate secure and transparent transactions opens up a world of possibilities for new forms of human interaction and collaboration. It can enable seamless and secure exchanges of value, information, and even ideas, fostering a more equitable and inclusive digital society.

As we navigate this path of technological advancement, the key lies in striking a delicate balance between the transformative power of blockchain and the preservation of our core human values. We must ensure that our digital future is not merely technologically enriched but also profoundly human, guided by the principles of empathy, compassion, and mutual respect.

Blockchain, when harnessed thoughtfully, can serve as a powerful tool for enhancing human connection and fostering a digital world that amplifies our collective human spirit. It can empower us to create a future where technology and humanity coexist in harmony, not as opposing forces but as mutually reinforcing partners in shaping a world of limitless possibilities.

This future, where blockchain's potential is fully realized, will not be one of technological dominance, but rather of human empowerment. It will be a future where the digital and human realms are not separate entities, but rather seamlessly intertwined, creating a tapestry of human existence that is more connected, more inclusive, and more profoundly human than ever before.

Embracing the Future: A Concluding Reflection on Humanity's Ledger

In this deep dive into blockchain's influence, we've gone beyond merely scratching the surface of its implications. This technology redefines the bedrock of our societal norms and reshapes professional identities. Through philosophical lenses, we've explored the complexities of trust in the digital age, uncovering how blockchain, with its unwavering ledger, both challenges and reinforces traditional notions of integrity and reliability.

Our sociological journey has transcended blockchain's technicalities, revealing its profound impact on social structures and personal interactions. We've witnessed the technology's role in revolutionizing the workplace, reshaping job roles, industries, and the very essence of professional relationships and organizational structures. This evolution extends far beyond economic transactions, touching the heart of our social interactions and the way we forge connections, collaborate, and create communal value.

Central to our exploration have been the ethical implications of embedding blockchain into our daily lives. We've engaged with the moral quandaries this technology presents, balancing the pursuit of innovation with the preservation of human values. The ethical considerations of blockchain's transparency and decentralization have been weighed against concerns of privacy, data ownership, and potential misuse.

As we look to the future, our insights guide our vision. We recognize that embracing blockchain is about more than technology adoption; it's about reshaping our modes of living, working, and interacting. Moving forward requires a balanced approach, one that synergizes blockchain's efficiencies with the depth of human connections and moral integrity.

We're poised not just to envision but to actively shape a future where blockchain is integrated thoughtfully into our society. This is a collective call to action for stakeholders across all spheres – technologists, policymakers, ethicists, and society at large – to collaboratively forge a world where technology uplifts humanity's highest ideals. Our aim is a future where blockchain not only revolutionizes our systems but also enriches and elevates the human experience, fostering a more connected, transparent, and equitable world.

A Future Forged by Technology and Trust

Our journey through the pages of this book has illuminated the transformative role of blockchain technology, transcending beyond mere transactions to profoundly impact societal structures and human relationships. We've seen how the certainties of technology intertwine with the complexities of human trust, creating a new tapestry for our interconnected world.

The Stewardship of Tomorrow:

The responsibility of shaping this future lies with each of us — technologists, policymakers, academics, and engaged citizens. Our collective decisions and actions in the coming years will define the integration of blockchain into the fabric of society. We are called to create an environment where technology is not just an instrument of progress but a catalyst for holistic well-being and ethical advancement.

Collective Action for a Conscious Future:

- Technologists are challenged to innovate with responsibility, ensuring advancements in blockchain and AI are anchored in ethical considerations and human-centric values.
- Policymakers and regulators must develop frameworks that harness the benefits of these technologies while protecting against potential misuses and ensuring societal welfare.
- Academics and thought leaders are encouraged to continue offering critical insights, challenging norms, and envisioning the evolving relationship between technology and society.
- Every individual has a role in this narrative. By staying informed, proactive, and engaged, we can collectively ensure that the digital world reflects our shared values and aspirations.

Through this exploration of blockchain technology

entwined with the essence of human trust, we've ventured into a realm abundant with philosophical insight, technological innovation, and sociological understanding. Our journey has revealed the unique intersection where blockchain's steadfast reliability meets the ever-changing landscape of human relationships and connections.

This narrative delves deeper than just the technicalities of blockchain or its potential uses. It's an exploration into how this technology can resonate with the core aspects of our humanity – our capacity for empathy, our boundless creativity, and our deep need for meaningful connections. It's about understanding that as our world becomes more entwined with technologies like artificial intelligence and automation, these advancements should not overshadow but rather enhance and celebrate the human experience. The true richness of our existence is found in our stories, creative expressions, and the bonds we forge with one another.

As we integrate technologies such as blockchain into our societal fabric, our focus must not waver from ensuring that these advancements enhance rather than eclipse our human essence. This journey transcends the mere creation of efficient and transparent systems; it's about forging an environment where technology serves as a guiding light, illuminating the path towards a brighter and more enriched human experience. We envision a future where technological progress fosters a sense of community, deepens our connections, and cultivates empathy and understanding. This is about navigating the evolving digital world with a profound appreciation for the values and experiences that are quintessentially human.

In concluding this book, I offer a collection of insights and reflections—a diverse array of thoughts and possibilities that have emerged from our exploration through time. This work is an invitation for each reader to continue this inquiry in their own lives, to challenge assumptions, to contemplate the implications, and to dream of a world where technology and humanity coalesce in perfect harmony. We pass this legacy on to you—a commitment to a future where our technological advances are more than mere conveniences; they are tools that uphold, celebrate, and amplify the essence of our shared human journey.

Let us all contribute to this future, creating a world where our digital advancements are not just functional tools but pathways to a more enriched, connected, and profoundly human experience. Together, we can shape a world where technology is not a force that distances us but a bridge that connects us, a tool that amplifies our collective human spirit, and a catalyst for a future where the boundaries between the physical and digital realms are not walls but seams, seamlessly woven into the tapestry of our existence.

References

Blockchain Technology Overview
- Nakamoto, Satoshi. "Bitcoin: A Peer-to-Peer Electronic Cash System." 2008. https://bitcoinwhitepaper.co/

Sociological Theories
- Durkheim, Émile. "The Division of Labor in Society." New York: Free Press, 1893.

- Weber, Max. "Economy and Society: An Outline of Interpretive Sociology." Berkeley: University of California Press, 1978.

Case Studies
- "Bext360 Case Study: How Blockchain Technology is Transforming the Real Estate Industry." https://www.techstars.com/newsroom/techstars-boulder-announces-b2b-fintech-thesis-for-the-july-2023-accelerator

- "IBM Blockchain Case Studies." https://www.ibm.com/blockchain/use-cases/

- "Estonia's E-Residency Program: A Case Study." https://www.e-resident.gov.ee/blog/posts/do-e-residents-feel-a-sense-of-belonging-to-estonia/

- "Blockchain and Land Registry: Lessons from Georgia and Sweden." https://eurasianet.org/georgia-authorities-use-blockchain-technology-for-developing-land-registry

DAOs (Decentralized Autonomous Organizations)

- "Aragon Whitepaper." https://cryptorating.eu/whitepapers/Aragon/Aragon%20Whitepaper.pdf

- "Moloch DAO: A Radical Experiment in Decentralized Governance." . https://molochdao.com/

- "The DAO: A Collaborative Governance Experiment." https://thedao.com/

Philosophical Perspectives on Trust

- Kant, Immanuel. "Groundwork of the Metaphysics of Morals." Cambridge: Cambridge University Press, 1998.

- Foucault, Michel. "The History of Sexuality, Volume 1: An Introduction." New York: Pantheon Books, 1978.

Ethical Implications of Blockchain

- "Blockchain and Ethics: A Discussion Paper." https://www.sciencedirect.com/topics/engineering/blockchain

- "The Ethical Implications of Blockchain Technology." https://medium.com/@MATCHAIN/exploring-the-ethical-implications-of-blockchain-technology-14709b55a30d

Blockchain in Social Justice

- "Blockchain for Social Impact: Moving Beyond the Hype." https://humanrights.wbcsd.org/project/blockchain-for-social-impact-moving-beyond-the-hype/

- "Blockchain and Social Justice: A New Frontier for Human Rights." https://www.un.org/en/un-chronicle/blockchain-and-sustainable-growth

Blockchain in Governance

- "Barcelona's Blockchain Strategy in Municipal Governance." https://smartcatalonia.gencat.cat/en/projectes/tecnologies/detalls/article/estrategia-blockchain-de-catalunya

- "How Blockchain Technology Can Transform Governance." https://blogs.worldbank.org/governance/blockchain-technology-has-potential-transform-government-first-we-need-build-trust

Decentralization and Security

- "Security Challenges in Decentralized Blockchain Networks." https://medium.com/@droomdroom/blockchain-security-safeguarding-your-assets-in-a-decentralized-world-5f55f485834c

- "Decentralization and Security: The Blockchain Trade-Off." http://essay.utwente.nl/91994/

Future of Work

- "The Impact of AI and Blockchain on the Future of Work."
  https://www.weforum.org/agenda/2023/08/ai-artificial-intelligence-changing-the-future-of-work-jobs/

- "The Future of Work in the Age of Blockchain."
  https://www.forbes.com/sites/emilyhe/2020/09/01/blockchains-role-in-reshaping-the-future-of-work/

- "The Future of Work and Society in the Age of AI." Aaron Vick.
  https://objkt.com/review/2022

Humanizing Technology

- "Balancing Blockchain Technology with Human Values."
  https://medium.com/@varunjay.varma/everyone-should-have-a-blockchain-df9b74ad4c16

NOTES:

Additional Reading

Books

1. Vick, Aaron. *Inevitable Revolutions: Secrets and Strategies for a Successful Business*. Leaderspress, 2021.

2. Vick, Aaron, et al. *QUITLESS: The Power of Persistence in Business and Life*. Leaders PR, 2021.

3. Vick, Aaron. "Part 1 - Passion" *7-Figure Minds: How to Grow and Lead a 7-Figure Business*. Edited by Alinka Rutkowska, Leaders Press, 2021.

4. Vick, Aaron. *Leaderpreneur: A Leadership Guide for Startups and Entrepreneurs*. AaronVick, 2020.

5. Vick, Aaron. *What is...Love is?* iUniverse, 2020.

Articles

1. Vick, Aaron. "Grinding the NFT community to death." **CryptoGlobe**. 5 March 2022. [Link](https://www.cryptoglobe.com/latest/2022/03/grinding-the-nft-community-to-death/)

2. Vick, Aaron. "How NFTs are creating social value." **Forbes**. 24 February 2022. [Link](https://www.forbes.com/sites/forbestechcouncil/2022/02/24/how-nfts-are-creating-social-value/)

3. Vick, Aaron. "Embracing Poaps & NFTs Beyond Personal Art Collections." **The AI Journal**. 12 February 2022. [Link](https://aijourn.com/embracing-poaps-nfts-beyond-personal-art-collections/)

4. Vick, Aaron. "Is Your Startup Secure? Here Are Seven Basic Cybersecurity Tips For Startups." **Forbes**. 24 September 2019. [Link](https://www.forbes.com/sites/forbestechcouncil/2019/09/24/is-your-startup-secure-here-are-seven-basic-cybersecurity-tips-for-startups/)

5. Vick, Aaron. "What Cybersecurity Trends Should We Expect From The Rest Of 2019?" **Forbes**. 20 August 2019. [Link](https://www.forbes.com/sites/forbestechcouncil/2019/08/20/what-cybersecurity-trends-should-we-expect-from-the-rest-of-2019/)

6. Vick, Aaron. "How Long Is My Startup Runway? A Guide To Calculating And Managing Monthly Burn Rate." **Forbes**. 31 July 2019. [Link](https://www.forbes.com/sites/forbestechcouncil/2019/07/31/how-long-is-my-startup-runway-a-guide-to-calculating-and-managing-monthly-burn-rate/)

www.ingramcontent.com/pod-product-compliance
Lightning Source LLC
Chambersburg PA
CBHW051247020426
42333CB00025B/3098